The R.A.M.S. Library of Alchemy

Volume 28

Lamspring's Process

Sigismond Bacstrom M.D.

Hans W. Nintzel and

Philip N. Wheeler

Editors.

R.A.M.S. Publishing Company

Lamspring's Process

Sigismond Bacstrom M.D.

Hans W. Nintzel and

Philip N. Wheeler,

Editors.

Produced by

Restorers of Alchemical Manuscripts Society

R.A.M.S. Publishing Company

R.A.M.S. Publishing Company
117 Rutherford Lane
Stuarts Draft VA 24477

Lamspring's Process
Copyright © 2015 R.A.M.S. Publishing Company

First Edition 2013
Second Edition 2015

ISBN-13 **978-1511637770**
ISBN-10 **1511637773**

Image Processing by Philip N. Wheeler

Printed in the United States of America

Table of Contents

Disclaimer 8

Introduction 9

Lamspring's Process 11

PREFACE 12

The Great Work of the Lapis Sophorum 15

    Figure I...................................19

    Figure II..................................23

    Figure III.................................27

    Figure IV..................................31

    Figure V...................................35

    Figure VI..................................39

    Figure VII.................................43

    Figure VIII................................47

    Figure IX..................................51

    Figure X...................................55

    Figure XI..................................59

    Figure XII.................................63

    Figure XIII................................67

    Figure XIV.................................71

    Figure XV..................................75

Sigismond Bacstrom's Commentary 79

INTRODUCTION 79

THE WORK 81

NOTES ON THE FOREGOING PROCESS 125

RECAPITULATION 128

Appendix A 148

GLOSSARY OF LATIN TERMS 151

A Word from the Publisher 155

Dedicated to Hans W. Nintzel,

American Alchemist

and

Founder of the

Restorers of Alchemical Manuscripts Society

(R.A.M.S.)

Disclaimer

Liability: The publisher does not warrant or assume any legal liability or responsibility for the accuracy, completeness, or usefulness of any information, apparatus, product, or process disclosed. The publisher makes no representation as to the accuracy or completeness of the contents of this book and specifically disclaims any implied warranty of merchantability or fitness for a particular purpose. No warranty may be created or extended by written sales materials or sales representatives. You should obtain professional consultation where appropriate. The publisher shall not be liable for any loss of profit or other commercial or personal damages, including but not limited to special, incidental, consequential, or other damages.

Introduction

Philip N. Wheeler

Lamspring's Process, written in Germany in the 15th century or earlier, was known to many of the ancient alchemists including Basil Valentine and Paracelsus. The original included fifteen enigmatic illustrations, included as grayscale images. The color illustrations in this edition are from the Latin translation of Nicolaus Majus. That manuscript from 1607 is titled Lambsprinck's *De lapide philosophico*. Many different sets of these illustrations have been created over the years.

In 1804, Sigismond Bacstrom translated the text from German, and that translation was used for this edition. He also added an extensive commentary, drawing upon commentaries from other alchemists as well as his own insight.

Dr. Sigismund Bacstrom (circa 1750-1805), born in Scandinavia, was a physician, alchemist, and Rosicrucian. In 1794 he was initiated into the Societas Rosae Crucis by Comte Louis de Chazal, an occultist and alchemist. Bacstrom met Chazal on the island of Mauritius during Bacstrom's travels as a ship's surgeon. This meeting led to Bacstrom's interest in alchemy. Eventually he settled in London where he practiced alchemy and other ancient arts. He avidly translated many alchemical works into English at the end of the 18th and the beginning of the 19th century.

Lamspring's Process

Sigismond Bacstrom M.D.

Translator and Commentator

Hans W. Nintzel

Philip N. Wheeler

Editors

PREFACE

Sigismond Bacstrom, M.D.

The subjoined processes for the manipulation of the Lapis Sophorum are those that were actually followed by a German nobleman, a great Philosopher and a real possessor of the name of Lamspring. He left behind him his process in hieroglyphical figures which were very well engraved on copper plates by Merian and published in a fourth treatise in 1625, which is very rare.

Whether Lamspring himself, or someone to whom he communicated his secrets, wrote the German original from which the following pages are translated is not known. The work is highly valuable regardless of its origin.

As this author gives plain instructions respecting the true Lac Virginis or mercurial water or oil of Paracelsus and other Philosophers, and stands highly recommended by Dr. Becher (Vide Stahl's *Chemistry* translated by Shaw p. 421 -27), and as the work upon Mercury per se, with a Solar or Lunar ferment, *in forma olei*, is the greatest of all mineral or metallic works, I do not wish that it should be lost with me. In case of my death, therefore, I have translated it for you from the German copy of the process which I have in my possession.

I shall only observe farther that Lamspring's fourth treatise above mentioned is written in a kind of Emblematical verses, which becomes perfectly intelligible, as do even the hieroglyphics themselves when the following pages are employed as a key.

The present work is perfectly intelligible and is free from all ambiguity, but the process, taken in all its parts (for the various manipulations described have but one ultimate object) is laborious and expensive, and demands an able operator. It appears from the Writings of Basil Valentine that he was acquainted with this work, as was also Paracelsus who is plainer, but not plain enough. Isaac Holland seems also to have known it. It was known also at the Court of Saxony by Prince Elector Augustus about the year 1580 to 1590, and by Rudolphus Secundus Emperor of Germany some few years after; and likewise by Christianus IV, son and successor to Augustus of Saxony, all which facts seem to be well attested by documents and writings which I have examined.

April 1804

The Great Work of the Lapis Sophorum

According to Lamspring's Process

I am called Lamspring, born of a Noble Family,
and this Crest I bear
with Glory and Justice.

Philosophy I have read, and thoroughly understood,
The utmost depth of my teachers' knowledge have I
sounded.
This God graciously granted to me,
Giving me a heart to understand wisdom.
Thus I became the Author of this Book,
And I have clearly set forth the whole matter,
That Rich and Poor might understand.
There is nothing like it upon earth;
Nor (God be praised) have I therein forgotten my
humble self.
I am acquainted with the only true foundation:
Therefore preserve this Book with care,
And take heed that you study it again and again.
Thus shall you receive and learn the truth,
And use this great gift of God for good ends.
O God the Father, which art of all the beginning and
end,
We beseech thee for the sake of our Lord Jesus
Christ
To enlighten our minds and thoughts,
That we may praise Thee without ceasing,
And accomplish this Book according to Thy will!
Direct Thou everything to a good end,
And preserve us through Thy great mercy. -
With the help of God I will shew you this Art,
And will not hide or veil the truth from you.
After that you understand me aright,

You will soon be free from the bonds of error.

For there is only one substance,

In which all the rest is hidden;

Therefore, keep a good heart.

Coction, time, and patience are what you need;

If you would enjoy the precious reward,

You must cheerfully give both time and labour.

For you must subject to gentle coction the seeds and the metals,

Day by day, during several weeks;

Thus in this one vile thing

You will discover and bring to perfection the whole work of Philosophy,

Which to most men appears impossible,

Though it is a convenient and easy task.

If we were to shew it to the outer world

We should be derided by men, women, and children.

Therefore be modest and secret,

And you will be left in peace and security.

Remember your duty towards your neighbour and your God,

Who gives this Art, and would have it concealed.

Now we will conclude the Preface,

That we may begin to describe the very Art,

And truly and plainly set it forth in figures,

Rendering thanks to the Creator of every creature.

Hereunto follows the First Figure.

Figure I

Be warned and understand truly that two fishes are swimming in our sea.

The Sea is the body, the two fishes are the Soul and Spirit.

The Sages will tell you
That two fishes are in our sea
Without any flesh or bones.
Let them be cooked in their own water;
Then they also will become a vast sea,
The vastness of which no man can describe.
Moreover, the Sages say
That the two fishes are only one, not two;
They are two, and nevertheless they are one,
Body, Spirit, and Soul.
Now, I tell you most truly,
Cook these three together,
That there may be a very large sea.
Cook the sulphur well with the sulphur,
And hold your tongue about it:
Conceal your knowledge to your own advantage,
And you shall be free from poverty.
Only let your discovery remain a close secret.

Figure II

Here you straightaway behold
a black beast in the forest.

Putrefaction.

The Sage says

That a wild beast is in the forest,

Whose skin is of the blackest dye.

If any man cut off his head,

His blackness will disappear,

And give place to a snowy white.

Understand well the meaning of this head:

The blackness is called the head of the Raven;

As soon as it disappears,
A white colour is straightway manifested;

It is given this name, despoiled of its head.

When the Beast's black hue has vanished in a black
smoke,

The Sages rejoice

From the bottom of their hearts;

But they keep it a close secret,

That no foolish man may know it.

Yet unto their Sons, in kindness of heart,

They partly reveal it in their writings;

And therefore let those who receive the gift

Enjoy it also in silence,

Since God would have it concealed.

Figure III

Hear without terror that in the forest are hidden
A Deer and a Unicorn.

In the Body there is Soul and Spirit.

The Sages say truly
That two animals are in this forest:
One glorious, beautiful, and swift,
A great and strong deer;
The other an unicorn.
They are concealed in the forest,
But happy shall that man be called
Who shall snare and capture them.
The Masters shew you here clearly
That in all places
These two animals wander about in forests
(But know that the forest is but one).
If we apply the parable to our Art,
We shall call the forest the Body.
That will be rightly and truly said.
The unicorn will be the Spirit at all times.
The deer desires no other name
But that of the Soul; which name no man shall take
away from it.
He that knows how to tame and master them by Art,
To couple them together,
And to lead them in and out of the forest,
May justly be called a Master.
For we rightly judge
That he has attained the golden flesh,
And may triumph everywhere;
Nay, he may bear rule over great Augustus.

Figure IV

Here you behold a Great Marvel
Two Lions are joined into one.

The Spirit and Soul must be united in their Body.

The Sages do faithfully teach us
That two strong lions, to wit, male and female,
Lurk in a dark and rugged valley.
These the Master must catch,
Though they are swift and fierce,
And of terrible and savage aspect.
He who, by wisdom and cunning,
Can snare and bind them,
And lead them into the same forest,
Of him it may be said with justice and truth
That he has merited the meed of praise before all
others,
And that his wisdom transcends that of the worldly
wise.

Figure V

A wolf and a dog are in one house,
And are afterwards changed into one.

The Body is mortified and rendered white, then joined to Soul and Spirit by being saturated with them.

Alexander writes from Persia

That a wolf and a dog are in this field,

Which, as the Sages say,

Are descended from the same stock,

But the wolf comes from the east,

And the dog from the west.

They are full of jealousy,

Fury, rage, and madness;

One kills the other,

And from them comes a great poison.

But when they are restored to life,

They are clearly shewn to be

The Great and Precious Medicine,

The most glorious Remedy upon earth,

Which refreshes and restores the Sages,

Who render thanks to God, and do praise Him.

Figure VI

This surely is a great miracle and without any
deception
That in a venomous dragon there should be a great
medicine.

The Mercury is precipitated or sublimed,
dissolved in its own proper water,
and then once more coagulated.

A savage Dragon lives in the forest,

Most venomous he is, yet lacking nothing:

When he sees the rays of the Sun and its bright

fire,

He scatters abroad his poison,

And flies upward so fiercely

That no living creature can stand before him,

Nor is even the Basilisk equal to him.

He who hath skill to slay him, wisely

Hath escaped from all dangers.

Yet all venom, and colours, are multiplied

In the hour of his death.

His venom becomes the great Medicine.

He quickly consumes his venom,

For he devours his poisonous tail.

All this is performed on his own body,

From which flows forth glorious Balm,

With all its miraculous virtues.

Hereat all the Sages do loudly rejoice.

Figure VII

We hear two birds in the forest,
Yet we must understand them to be only one.

The Mercury having been often sublimed, is at length fixed, and becomes capable of resisting fire: the sublimation must be repeated until at length the fixation is attained.

A nest is found in the forest,

In which Hermes has his brood;

One fledgling always strives to fly upward,

The other rejoices to sit quietly in the nest;

Yet neither can get away from the other.

The one that is below holds the one that is above,
And will not let it get away from the nest,

As a husband in a house with his wife,

Bound together in closest bonds of wedlock.

So also do we rejoice at all times,

That we hold the female eagle fast in this way,

And we render thanks to God the Father.

Figure VIII

Here are two birds, great and strong;
The body and spirit; one devours the other.

Let the Body be placed in horse-dung, or a warm
bath, the Spirit having been extracted from it. The
Body has become white by the process, the Spirit red
by our Art. All that exists tends towards
perfection, and thus is the Philosopher's Stone
prepared.

In India there is a most pleasant wood,

In which two birds are bound together.

One is of a snowy white; the other is red.

They bite each other, and one is slain

And devoured by the other.

Then both are changed into white doves,

And of the Dove is born a Phoenix,

Which has left behind blackness and foul death,

And has regained a more glorious life.

This power was given it by God Himself,

That it might live eternally, and never die.

It gives us wealth, it preserves our life,

And with it we may work great miracles,

As also the true Philosophers do plainly inform us.

Figure IX

The Lord of the Forests has recovered his kingdom, and mounted from the lowest to the highest degree. If fortune smile, you may from a rhetor become a consul; if fortune frown, the consul may become a rhetor.

Thus you may know that the Tincture has truly attained the first degree.

Now hear of a wonderful deed,

For I will teach you great things,

How the King rises high above all his race;

And hear also what the noble lord of the forest

says:

I have overcome and vanquished my foes,

I have trodden the venomous Dragon under foot,

I am a great and glorious King in the earth.

There is none greater than I,

Child either of the Artist or of Nature,

Among all living creatures.

I do all that man can desire,

I give power and lasting health,

Also gold, silver, gems, and precious stones,

And the panacea for great and small diseases.

Yet at first I was of ignoble birth,

Till I was set in a high place.

To reach this lofty summit

Was given me by God and Nature.

Thence from the meanest I became the highest,

And mounted to the most glorious throne,

And to the state of royal sovereignty:

Therefore Hermes has called me the Lord of the

Forests.

Figure X

A salamander lives in the fire, which imparts to it
a most glorious hue.

This is the reiteration, gradation, and amelioration
of the Tincture, or Philosopher's Stone; and the
whole is called its Augmentation.

In all fables we are told
That the Salamander is born in the fire;
In the fire it has that food and life
Which Nature herself has assigned to it.
It dwells in a great mountain
Which is encompassed by many flames,
And one of these is ever smaller than another -
Herein the Salamander bathes.
The third is greater, the fourth brighter than the
rest -
In all these the Salamander washes, and is purified.
Then he hides him to his cave,
But on the way is caught and pierced
So that it dies, and yields up its life with its
blood.
But this, too, happens for its good:
For from its blood it wins immortal life,
And then death has no more power over it.
Its blood is the most precious Medicine upon earth,
The same has not its like in the world.
For this blood drives away all disease
In the bodies of metals,
Of men, and of beasts.
From it the Sages derive their science,
And through it they attain the Heavenly Gift,
Which is called the Philosopher's Stone,
Possessing the power of the whole world.
This gift the Sages impart to us with loving hearts,
That we may remember them for ever.

Figure XI

The father and the son have linked their hands with those of the guide: know that the three are body, soul and spirit.

Here is an old father of Israel,

Who has an only Son,

A Son whom he loves with all his heart.

With sorrow he prescribes sorrow to him.

He commits him to a guide,

Who is to conduct him whithersoever he will.

The Guide addresses the Son in these words:

Come hither! I will conduct thee everywhere,

To the summit of the loftiest mountain,

That thou mayest understand all wisdom,

That thou mayest behold the greatness of the earth,

and of the sea,

And then derive true pleasure.

I will bear thee through the air
To the gates of highest heaven.

The Son hearkened to the words of the Guide,

And ascended upward with him;

There saw he the heavenly throne,

That was beyond measure glorious.

When he had beheld these things,

He remembered his Father with sighing,

Pitied the great sorrow of his Father,

And said: I will return to his breast.

Figure XII

Another mountain of India lies in the vessel, which
the spirit and the soul – that is, the son and the
guide –
have climbed.

Says the Son to the Guide:

I will go down to my Father,

For he cannot live without me.

He sighs and calls aloud for me.

And the Guide makes answer to the Son:

I will not let thee go alone;

From thy Father's bosom I brought thee forth,

I will also take thee back again,

That he may rejoice again and live.

This strength will we give unto him.

So both arose without delay,

And returned to the Father's house.

When the Father saw his Son coming,

He cried aloud, and said:

Figure XIII

Here the father devours the son;
The soul and spirit flow forth from the body.

My Son, I was dead without thee,

And lived in great danger of my life.

I revive at thy return,

And it fills my breast with joy.

But when the Son entered the Father's house,

The Father took him to his heart,

And swallowed him out of excessive joy,

And that with his own mouth.

The great exertion makes the Father sweat.

Figure XIV

Here the father sweats profusely, while oil and the
true tincture of the sages flow from him.

Here the Father sweats on account of the Son,

And earnestly beseeches God,

Who has created everything in His hands,

Who creates, and has created all things,

To bring forth his Son from his body,

And to restore him to his former life.

God hearkens to his prayers,

And bids the Father lie down and sleep.

Then God sends down rain from heaven

To the earth from the shining stars.

It was a fertilizing, silver rain,

Which bedewed and softened the Father's Body.

Succour us, Lord, at the end,

That we may obtain Thy gracious Gift!

Figure XV

Here father and son are joined in one so to remain forever.

The sleeping Father is here changed
Entirely into limpid water,
And by virtue of this water alone
The good work is accomplished.
There is now a glorified and beautiful Father,
And he brings forth a new Son.
The Son ever remains in the Father,
And the Father in the Son.
Thus in divers things
They produce untold, precious fruit.
They perish never more,
And laugh at death.
By the grace of God they abide forever,
The Father and the Son, triumphing gloriously
In the splendour of their new Kingdom.
Upon one throne they sit,
And the face of the Ancient Master
Is straightway seen between them:
He is arrayed in a crimson robe.

TO THE INVISIBLE KING OF THE WORLD,

TO THE ONLY TRUE AND IMMORTAL GOD

BE PRAISE AND GLORY

NOW AND EVERMORE.

AMEN.

Sigismond Bacstrom's Commentary

INTRODUCTION

He that knows how to elaborate the Great work from Mercury alone will be the most profound Indicator of science and Art! because in mercury alone is to be found what the Wise Masters look for "ist in mercurie quidquid quaerunt Sapientes.[1]"

Mercury, understand me literally, is the mother of all the ductile Metals the femine sperm, the body, the menstruum and N.B.[2] the nearest matter.

Mercury is not only a Spiritual Essence but also an Essential Body — a Natura Media, containing a tinging sulphur and mercury indeterminated. Mercury dies and resuscitates, and is fixed by its own depurated elements.

N.B. But it is highly necessary that Mercury be depurated of its impurities, which are Earth and Water--the two passive elements. (Vide Philosophical Canons, which teaches the same doctrine).

---

[1] Whatever wise men seek is in mercury. -pnw
[2] Nota bene, meaning Note well. -pnw

Mercury may be extracted out of Metals, and out of some Marcasites, such as Antimony, Zinc, Bismuth, Red golden ore and, as likewise from common running mercury itself. They are all of a nearly similar nature, excepting only that the mercury of Gold, Antimony, Zinc, Iron, and Copper is of a solar nature, and that of Bismuth, red golden (colored) ore, Lead, Tin, and Silver is of a Lunar nature. You may prepare the Lapis from a lunar as well as a solar, or from an indeterminated mercury, if your mercury be but highly pure, so that you have the mercury out of the mercury. Then with a lunar or a solar spiritualized ferment you may lead your mercury which may you please; and even if you use a solar mercury you must absolutely make it pass through the perfect Lunar White Tincture before you can possibly obtain the Solar Red Tincture. Thus you may find our pure mercury and sulphur above ground, composed of the selfsame elements as those which in the mines generate silver and Gold.

THE WORK

PURIFICATION OF THE MERCURY

Take good Spanish Mercury, or that from Istria in Italy, one or two pounds. Rub it in a wooden, or in a porphyry mortar, with sea salt and sharp vinegar,

until the salt ⚹ vinegar become black. Then wash the mercury with water. Continue thus to rub with more salt and vinegar until the greater part of the external Filth is gone. Wash it again, dry it and then strain the cleansed mercury through chamois leather. The mercury being thoroughly dry should be passed repeatedly through the leather till it appears very bright and beautiful.

SUBLIMATION OF THE ☿

Rub and unite your cleansed mercury with an equal weight of good mercury sublimate corrosive, in a porphyry mortar, until it becomes a grey mass. Weigh the mass and mix it with an equal weight of pure nitre[3] and roman vitriol, of each $\overline{aa}$ (equal parts). Rub all well together until the ingredients are well incorporated and appear like a paste.

Put the paste into a strong subliming glass, whereon place an alembic, leaving the pipe open for the evaporation of the humidity. Place your subliming body pretty deep in sand; increase your fire under the sand gradually, and sublime all the mercury upwards into the alembic, white as snow. When you observe this taking place shut the pipe of the alembic.

This sublimation must be repeated three times: that is, once with the ingredients and twice after per se, in order to obtain the ☿ (sublimated mercury) as pure as possible.

---

[3] Dr. Bacstrom has here overlooked one circumstance. The mercury was sublimed with corrosive sublimate and that contains the acid of sea salt. Of nitre, aqua fortis is the only acid. —hwn

HUMID CALCINATION OF THE ☿

Take one pound of this ☿ finely powdered, and put it into two pounds of good Aqua fortis, not all at once, but gradually, in a glass body. Project only 2 ℥ (ounce) at once into the aqua fortis, and so proceed until the 16 ℥ are all dissolved, as sugar dissolves in wine. Shut the glass close, and place it in a warm balneum, in such a heat that the glass may feel pleasantly warm, but not hot, and let it stand so for ten days to insure a true solution per minima.

Now apply an alembic to the glass body, and with a little more heat, in a hotter balneum, distill the aqua fortis from the mercury into a luted receiver, and the mercury will be left at the bottom of the glass body white like Hogs lard: then cease distilling. This is the true philosophical humid calcination of the mercury.

EXUBERATION OF THE ABOVE ☿

The mercury that has undergone the humid calcination must be exuberated and rendered fusible, which is accomplished in the following manner:
Cover the bottom of your glass body with a good tough luting. Lute on a roomy alembic and leave the pipe open. Place your body pretty deep in sifted ashes or very fine sand. Increase your fire under the iron pot gradually until the humidity is all gone, and then again till you have sublimed your mercury into the alembic. Beware of the poisonous invisible fumes, which are mortal when received into the lungs by inspiration.

When no more ascends let the fire die away, and leave the vessel in its place to cool.

Next morning (defending your mouth and nose with a towel, moistened with good vinegar) take off the alembic, take out your sublimate and put it again into the glass body, previously cleansed, washed and dried, or put it into a new one. Then proceed as before and sublime the mercury per se. Repeat the sublimation per se once more (that is three sublimations in all) and the sublimate will appear of a most brilliant glittering white, and will be much more visible than before. This is our

☿exuberatus, or fusibilis.[4] (♄♐ have this Dryway.)

---

[4] What the Author says respecting sea salt and nitre is perfectly just, but in the first instance he had sublimed his mercury with nitre and vitriol in equal part. Perhaps nitre there was a mistake for sea salt; for we have enough of nitre in the Aqua fortis in which he dissolves his first sublimate, but whence have we sea salt? It stands however in my original as I have written it, nor do I believe that the first sublimation being done with nitre and Roman Vitriol would be inferior to that with sea salt and vitriol. In fact, we have a number of processes that proceed either way.

FIXATION OF THE FOREGOING EXUBERATED ☿, THAT IT MAY BECOME THE GLUTEN AQUILAE OF PARACELSUS.

Of your foregoing Exuberated 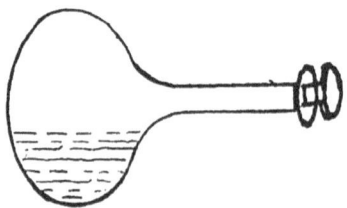 you ought to have one pound and a half at least prepared. Nor will you have cause to grudge your labour.

Take of your Exuberated sublimated Mercury, which has now lost its internal humidity and external earth by the sublimations, a third part, that is half a pound, put it into a glass of this form:

which must have a glass stopper, and ought to be pretty strong and roomy enough to be able to admit of subliming and fixing the half pound of your . It ought to be blown with a flattish bottom, without any knob.

This glass is not to be placed on its bottom but sideways, with that part of the belly which contains

the sublimate buried in fine sand, in an iron pan,
over a charcoal furnace, in this manner.

Increase your heat gradually, and continue until the
sublimate has ascended into the upper side of the
bottle; then let the fire die away.

Next morning turn the upper side down into the sand
and bury in the sand that part which contains the

sublimate. Light the fire ♂ sublime again -- and
so continue to do every day for a fortnight or three
weeks, until your Sublimate, even in a very strong
heat will ascend no more, but remains below, fixed
and fusible.

Or you may fix your sublimated mercury in a double
glass, like two deep cups in this form.

The mouth of the one cup must be nicely filled and ground into the other. This vessel as well as the former one (whichever of them is made use of) must be heated to expel the cold air and all humidity,

before the previously warmed ♀ is introduced. Then the joining, or the stopper if the first mentioned vessel is used, must be luted outside with stripes of linen pasted over the joining five or six fold.

You could likewise fix your ⏜ (sublimate) in such a double glass as has been just described by applying the heat to the upper one, per ignen suppressionis: but one of the former methods is less troublesome and should therefore be preferred.

When you have proceeded thus far you are in possession of the Gluten Aquilae of Paracelsus, or

the first fusible body, or the fixed exuberated
earth of the mercury, perfectly pure!

Preserve it carefully for future use.

SOLUTION OF A PORTION OF THE RESERVED EXUBERATED MERCURY TO CONVERT IT INTO THE TRUE MERCURIAL WATER, WHICH IS THE TRUE LAC VIRGINIS OF THE SOPHI.

Take one half of your reserved two parts, that is half of a pound of your sublimed exuberated ☿, put it into a digesting glass ♇ shut it slightly, so as only to keep out the dust and humidity.

Place this in a luke-warm water bath, wherein you can constantly bear your hand, about blood hot; and let it continue there day and night, until your Mercury sublimate is all dissolved into a fat water, which will infallibly happen in a few days.

Or

Lay it on a strong glass plate or a slab of porphyry placed obliquely in a dry cool cellar, so that the air may strike over it, placing a glass funnel in a bottle under the lower end. The exuberated mercury sublimate will run per delequium into a water. This must be dephlegmated in a Balneum Mare: the first method does not require that.

Or

Hang your exuberated sublimated mercury in a strong, new linen, sharp pointed bag, over a glass funnel in a cool dry cellar, where there is a draught of air, and it will flow by attraction and drop into the funnel and bottle. This also must be deflegmated in a B. M.[5]

When finished put it into a strong glass bottle, with a glass stopper, and keep it for use.
This is the true Lac Virginis, or Mercurial Water -- the Dragon that devours and fixes its own tail, as Lamspring mentions, but it wants another apuration which is as follows:

---

[5] Balneam Mariae, Bath of Mary. A water bath resembling a double boiler. -pnw

DISTILLATION AND PURIFICATION OF THE LAC VIRGINIS WHICH YOU HAVE OBTAINED

Put the mercurial water which you have obtained into a digesting glass, which shut close. Then set it to digest in a gentle heat, not above 90 degrees, over a lamp for nine or ten days to procure a more intimate union between the sea salt and the mercury: and note here diligently, that in sea salt the universal mercury lies concealed, and in nitre the universal sulphur of nature, unspecificated.

After the digestion pour it cold into a glass body, apply an alembic, lute well the joining of the alembic and receiver, and distill the contents over in a B.M. or in an equally gentle and well managed heat in ashes, and you will obtain a pure mercurial water or Lac Virginis, perfectly homogeneous, as being the Mercury of Mercury, or The very mercurial volatile Essence thereof -- the Spirit of the White Mercurial stone, or the Spirit of the white Tincture of mercury, (With which spirit the author afterwards multiplies the Stone in Power and Virtue) and this is new, the true and genuine Lac Virginis, or aqua mercurii of Lamspring and other philosophers that have proceeded in this way.

CONJUNCTIE SPIRITUS CUM CORPORE
THE UNION OF THE SPIRIT WITH THE BODY OF MERCURY

Now take of your reserved fixed Gluten Aquilae one
part (one $\mathcal{Z}$ or as much as you like and can afford)
rub it into a warmed, perfectly dry glass mortar to
a subtil powder, which put into a conveniently sized
digesting globular Glass: pour upon it an equal
weight of its own Mercurial spirit or Lac Virginis:
do this gradually and when all is in shut the glass
immediately and lute the stopper.
Now you have united Man and Wife, the fixt with the
volatile, the body with the spirit, the salt with
the Mercury -- the sulphur being contained in the
Mercury.

This is Lamspring's Dragon which is going to devour
its own tail.[6]

---

[6] its own fixt salt, or Gluten Aquilae —hwn
"The Dragon is going to devour its own tail". The volatile,
the female, predominates at the first in every process for the
Lapis, as the first woman committed sin and disobedience
first. The female, the Virgin Mary also prevailed in
manifesting and corporifying the Messiah. She was the Material
Instrument to manifest the Messiah, who has restored that
which was lost by the first woman. You may perceive by this
also that the Male agent must spiritually and materially
prevail at last, and purify and fix the female, the volatile,
into one united glorified homogeneous and immortal essence, no
longer susceptible of a distinction of the sexes, for which
reason there can be no distinction of sexes after physical
death-—all must then be male agents, perpetually active

DIGESTION

Place your Globular Glass in a blood warm balneum. You should be able always to bear to have your hand in the warm water. This will be a guide to you as to the heat (A thermometer will, however, be more certain).

Let it stand to die and putrefy, for the space of 150 days, or five months; and be careful not to move nor disturb your glass.

When forty days have passed the first blackness will appear, which is called Caput Corvi, the Crows head, this will continue for some time.

After the blackness various intermediate colors will be seen, and lastly and gradually it will become white.

Then by increasing the heat of the bath a little, about 20 or 25 degrees, the White matter will ascend, and hang round the sides of the globe, and assume the appearance of fishes eyes. It will however settle again and look Silver White. This is new Sulphur naturae album, or the white sulphur of

unchangeable. -Bacstrom

nature, indeterminate. But should it remain fixed on the side of the glass it is equally useful.

SOLUTION OF THE WHITE CORPOREAL SULPHUR OF NATURE

Weigh this Sulphur naturae album, put it into a small glass body, and pour upon it double its weight of genuine rectified spirit of wine. Shut the glass close, with a blind alembic and digest it eight days in a blood-warm balneum. Then apply a proper alembic and receiver and, with a little more heat, distill the spirit of wine from the White sulphur of nature, till what remains behind looks like a white oil. As soon as you observe this oil you must cease.

Now you have prepared the oil of the White sulphur naturae, out of mercury alone.

This must be united with the White Ferment, with the Liquor or Oleum Sulphuris Lunae, which we shall teach you hereafter.

This is the surest and the most certain process to elaborate the Lapis Philosophorum via humida from Mercury and Spiritualized Ferment; and this is Lamspring's way; and believe me, the Great Elixir cannot be made without adding the ferment of Silver to the White Sulphur of Nature, and the Ferment of

Gold to the more digested red ☿ of Nature!

The Sulphur of Silver must be dissolved in the same manner as the sulphur naturae album made of Mercury. The Oil of the Sulphur Lunae is Anima or Ferment spiritualized, which must be united with the spirit and body of the Mercury; and this is called the first Spiritual fermentation of the Lapis, that it may become a Powerful Elixir. This vivifies the lapis.

The first union was only a union of the spirit and body of the mercury, but this second union is a threefold copulation of the anima (oleum sulphuris lunae), with the Spirit (lac virginis), and the body (gluten aquilae). The two last, viz., the Spirit and body, you have in the oleum sulphuris naturae albi, in a regenerated spiritualized state, agreeable to the emblematical figure in Lamspring's printed Treatise.

THE SPIRITUAL FERMENTATION OF THE SULPHUR NATURAE ALBUM OR THE INDETERMINATED WHITE TINCTURE

Take one part (⚴ or ⚴) of your oleum sulphuris lunae, put it into a warmed digesting globe, and add

to it three parts (⚴ or ⚴) of your oleum sulphuris naturae albi, that is, the dissolved white Sulphur of Nature reduced to an Oil, which Sulphur you made of mercury. Shut the glass, and after the superfluous humidity is gone[7], lute the glass stopper.

Place your glass in sifted ashes, in a gentle degree of dry heat, from 90 to 100 degrees. Let it continue till it dries into perfect permanent whiteness. Before this happens various transient colours will pass: then the matter will become white and glittering like fine silver.

This will only require a few weeks and you have then the Lapis Albus, or White Tincture completed, which tinges all metals (excepting gold) into fine silver; but one part will only transmute ten and no more. It may, however, be increased in power.

---

[7] after 24 hours —hwn

MULTIPLICATION OF THE LAPIS ALBUS OR
WHITE TINCTURE IN QUALITY OR VIRTUE

Having prepared the White Tincture you ought to multiply it in virtue, strength and power; otherwise your advantage will be but small.

Therefore, dissolve your White Tincture in your rectified Lac Virginis made of mercury, and when it is perfectly dissolved, distill the Spiritual liquor from it gently, until there remains a fixed oil of a white colour.

This fixt oil you must coagulate and dry up, in a globe or digesting glass set in ashes in a heat of 100 or perhaps 120 degrees.

This Solution and Exsiccation must be repeated three or four times more, until it will no more dry up but remains a fixed incombustible oil.

You have now obtained the White Tinctural Oil, or Great White Elixir of the Higher Order, by us called Tertiae Ordinis.

MULTIPLICATION IN QUANTITY

The multiplication in quantity, by simple projection, is performed in the following manner:

Take one part (ζ or ζ) of your white Tinctural oil or White Elixir tertiae ordinis, and project it upon

100 parts (ζ or ζ) of fine copelled silver in fusion in a crucible, and let them flow together for a full hour. All is safe now. You can hurt nothing by taking time enough. Make a trial by dipping a clean iron rod into it: examine the adherent matter, which ought to be and will be a brittle, white, vitrious mass. Then it has been long enough to be perfect, but should it remain exposed never so long to the heat you can no longer destroy it.

When the mass is cold beat it to a fine powder. This is the corporeal fermented white metallic Elixir.

Take one part (ζ or ζ) of this glassy powder and put this to one hundred parts of purified mercury in a crucible. Give it a good heat and let stand for an

hour in the fire in a wind furnace. The mercury will
not fly away, but will be converted into a fixt,
fusible, white, lunar, tinging precipitate.

PROJECTION

Now take 1 $\frac{z}{3}$ of this your last made Mercurial Lunar precipitate, envelope it in wax, and project it upon 100 ounces of common mercury, lead or tin, and let

---

[8] Respecting Projection observe the astonishing extensibility of the Elixir tertiae ordinis, which before multiplication transmuted only 10 parts of the inferior metals into silver but after multiplication can transmute at least 100 parts. The reason of this wonderful extensibility and Penetration is owing to the first preparation of the Sulphur Naturae album, which is a regenerated, resuscitated, indeterminated Essence. This again is fermented not corporeally only in the crucible, but spiritually, by a regenerated, spiritualized fermentum Lunae, reduced to an oil and then fixed together. Then the Elixir tertiae ordinis vitrifies silver, which again converts mercury into a tinging precipitate before it can become ductile silver in fresh mercury.

Afterwards, when you have obtained the perfect red sulphur naturae ex mercurio, fermented it spiritually with regenerated solar ferment reduced to a fixt oil, and multiplied it by Lac Virginis, the product is still more extensible in as much as the red sulphur naturae is or consists entirely of corporified fire, which causes it to be of a fiery red color, and as gold in its own original character is far more extensible than silver. The produce in gold, therefore, is immense, and if you reserve but a small portion of it is still farther multipliable, as Irenaeus, Count Bernardus, Basilius and all true Philosophers attest. This process of Lamspring is founded on true natural principles and is highly valuable.

them melt well together, in a good strong heat, for half an hour, and your mercury, lead or tin will be converted into most pure Silver.

Should it prove brittle or fly under the hammer, you must gradually add a little more of the same metal, ☿, ♄, or ♃, until your metal becomes soft and ductile fine silver -- finer than any from the Spanish mines in America.[8]

HOW TO PREPARE THE LUNAR FERMENT, AND TO SPIRITUALIZE IT INTO THE TRUE OLEUM LUNAE TO DETERMINATE THEREWITH THE SULPHUR NATURAE ALBUM EX MERCURIE TOWARDS LUNA

What you sow you will reap.

Gold produces a Solar and Silver a Lunar Tincture. Whosoever knows how to tinge Sulphur naturae indeterminatum, the White with Silver and the Red with Gold, will obtain the highest and most glorious secret in nature!

Hermes says: "Our Elixir is nothing else but Mercury fermented with Silver or with Gold." By his ☿ he means Sulphur naturae album et rubrum by Silver and Gold, he means, the Spiritual Lunar and Solar ferment. Both united (Sulphur naturae album with oleum lunae -- Sulphur naturae rubeum with oleum Solaris) constitute the true Mercurius Sophorum animatus, duplex or duplicatus which absolutely can become nothing else but the Great Elixir or Lapis Philosophorum.

CALCINATION OF THE SILVER

Take four, five or six ounces of fine copelled silver: beware that there be no copper in it, left by a careless Refiner. Let this be milled first and beat out into leaves at the Gold beater's: or make your Silver into a fine calx or Luna cornea, which edulcorate thoroughly with warm water and then dry it.

Take 4 or 5 ℥ of this calx or of your Silver leaf: and pour twice the weight of our rectified mercurial water or Lac Virginis into a digesting glass and dissolve therein, gradually, your calx or silver leaves, two or three at a time.

If you should be short of your Aqua Mercurialis you may mix it with $\overline{aa}$ clear Aqua fortis and the effect, as I have found by experience, will be the same.

Note! that you dissolve your silver at the first without heat; but when it will no longer act, being nearly loaded with ☽, then put your digesting glass in a blood warm water bath, and let it dissolve as much as it can -- You
have then obtained the true proportion.

Let the Glass be closely shut, and let it stand in the warm water bath nine or ten days, until the whole solution of Silver has become a green colored water. Then let the balneum cool. Take the glass out of the water, not quite cold, pour the solution carefully into a glass body not too high. If there happens to be a little sediment leave that carefully behind Apply quickly an Alembic and lute it to the body. Apply a receiver which also lute to the pipe of the alembic. letting the pipe go deep through the neck into the body of the receiver.

Place the body in a balneum and distill over gently the dissolving water into the receiver, until the greater part is come over. Do not hurry but proceed gently, until your silver remains behind in the body, not as a calx, but in the form of a white oil or oily liquid. Then take away the fire quickly and cease.

Our Lac Virginis or Aqua mercurii, either alone or when mixed with an $\overline{aa}$ of good Aqua fortis, is such a powerful solvent that nothing can resist or withstand it. It dissolves everything, for which reason the Philosophers before me have, not unjustly, called it Alcahest, universal dissolvent and Ignis Gehennae but we call it Our calcining water.[9]

It is of such a fiery nature ♂ property that it dissolves all metallic bodies into a liquid, which elementary fire cannot do, but reduces them into calces while this our calcining water reduces them into a metallic oil.[10]

---

[9] Some particulars respecting the Alkahest may be seen in Van Helmont. See also Boerhave's Chemistry translated by Shaw to Vol. 1, page 570 ♂ .

[10] See Isaacus Hollandus.

FURTHER SOLUTION AND SUBTILISATION OF THE OILY LIQUOR OF SILVER.

In order to subtilise this Liquor lunae still further and to deprive it of the corrosive moist fire, pour the oily liquor lunae into another glass fit for digestion.

Now you must have at hand some highly rectified Spirit of Wine, made from good German or French wine brandy, and not from corn: if from corn you will be deceived.

Pour, very carefully, a small quantity, a coffee spoonful at a time, into the Lunar liquid, and move the glass. When the two fires meet a great reaction takes place, and the glass gradually becomes intensely hot: therefore, you must proceed gradually and cautiously, shaking the glass horizontally, after each addition of the Spirit of wine, until it cools again and tranquility is restored. Continue thus adding the Spirit of Wine gradually until it stands four fingers breadth above the liquor lunae. By attending to these precautions you accomplish the union without any accident.

Then shut the digesting glass, which ought to have a long neck, and set it in a blood warm water bath, where let it remain to digest for ten days, until a perfect union has been effected, in form of a delicate oil liquid.[11]

Pour the oily liquid you have obtained by the digestion in balneo into a glass body; apply an alembic and with a most gentle heat, distill all the Spirit of wine from the dissolved silver, until the Silver remains again in the form of an Oil.

---

[11] See Appendix A.

FURTHER SUBTILISATION, AND PUTREFACTION OF YOUR OLEUM LUNAE

Put your Oleum or Liquor Lunae into a digesting globe of such a size that only one third or one fourth part of it may be filled. Then place the glass in warm water bath, so that it may constantly feel comfortably warm, as feel yourself when in perfect health.

Let the superfluous humidity evaporate during the first twenty four hours, then shut the glass tight with its glass stopper and luting.

Let it stand unmoved one hundred and fifty days, i.e. about five months, and the Silver will die and putrefy. When the caput corvi or blackness is past increase the heat twenty or thirty degrees and various beautiful colors, like the Peacocks tail, but transient, will pass from day to day, in the same manner as during the first regeneration of mercury, before taught.

After six weeks more you will see the White Sulphur appear. Then increase your heat again a little more and your White Sulphur will ascend and settle all round the sides of the globes, bright and shining,

like Fishes eyes. This is your purified,
resuscitated and regenerated Sulphur or Fermentum
Lunae.

This firmentum lunae is not so firm as a fixed body,
not is it so volatile as a spirit. It is a Natura
media, between the body and the spirit, and is
called the Sophic ferment, the Forma of the White
Elixir spiritualized, of a middle nature.
Without this Form the Lapis cannot tinge into
silver.

Note! that with this Sulphur Lunae the white
sulphurs of all the imperfect metals, can be
spiritually fermented, and become Tinging medicines,
which tinctures, when dissolved and coagulated three

times, in such a manner that they remain ☿ stand
like a fixed oil, are then incombustible oils, and
Elixeria tertia ordinis, as well as ours of Mercury,
and are equally multipliable.

REDUCTION OF THE LUNAR FERMENT OR SOPHIC FERMENT, SHINING LIKE FISHES EYES, INTO AN OIL

Take your Fermentum Lunae out of the digesting globe, put it into a clean digesting glass and pour highly rectified Spirit of wine (made from good brandy) upon it, so as to cover it two or three fingers breadth, and then digest in a blood warm balneum for two or three days.

After this digestion distill the Spirit of Wine gently from it until the lunar ferment remains behind in the form of an oil.

This is the Oleum Lunae for the Spiritual fermentation of the Sulphur naturae Ibum ex Mercurio, which you have also, by means of spirit of wine reduced to an oil; and thus the two oils are united per minima, as we have taught you, and want only to be dried up and fixed.

OF THE RUBIFICATION OF THE WHITE SULPHUR NATURAE EX MERCURIO

Having obtained the White Sulphur of Nature from mercury, in two or three digesting globe glasses, take that glass which you propose to continue to digest till it be perfected into the Red Sulphur, and, without permitting it to cool, place it in a Lamp furnace, in a bed of sifted ashes, warmed to the same degree of heat as the glass had acquired in the water bath. The dry heat in ashes must be no stronger than that you can bear the glass in your open hand.

Continue this gentle degree of dry heat, say about 120 to 130 degrees until your Sulphur naturae album has become of a very bright and beautiful cinnabar color, which it will in about thirty days.

This is Sulphur Rubrum Naturae indeterminatum.

SOLUTION OF THE RED SULPHUR NATURAE INTO AN OIL

Dissolve this Red sulphur of Nature by the same process as you did the White Sulphur: that is dissolve it in genuine, highly rectified Spirit of Wine, digest in a blood warm water bath, keeping the glass close shut, and you will obtain a deep, Ruby-red, transparent solution.

This Solution is Fire!

If you tinge a bottle of good old White Rhine wine or Austrian wine with this Essence until the same becomes as deep in color as Burgundy, which a small quantity of the dissolved red Sulphur will effect, you have then in your possession.

THE GLORIOUS UNIVERSAL MEDICINE, OR QUINTA ESSENTIA MEDICINALIS

Which is so powerful that a few doses of a coffee spoonful will expel the most dreadful diseases Epilepsy, palsy, dropsy, consumptions, fevers, gout, leprosy, all fly before it. It is a cure for the maladies of the whole animal creation.

But when the Solar sulphur spiritualized, has been united, and coagulated therewith, it then becomes a hundred times more powerful, and must therefore be diluted proportionally before it be exhibited as a medicine. One single grain in substance, in that state would extinguish life like a stroke of lightning or a violent shock of Electricity which is the same thing with less power, as we have proved by experiments made on dogs and other animals.

DISTILLATION FOR THE RUBY-RED TRANSPARENT SOLUTION OF THE RED SULPHUR OF NATURE.

Having by the means directed obtained your ruby-red transparent solution of the Red sulphur of Nature in Spirit of wine, you must, with a gentle heat in Balneo, draw off the spirit of wine Per Alembicum, until there remains behind a Ruby-red Oil.

COMPOSITION OF THE PRINCIPLES

To three parts of the Ruby colored oil you must add one part of the Golden ferment reduced to an Oil, by means of Spirit of Wine.

Manage exactly as you did the White, and coagulate the united oils in a digesting globe glass, placed in a dry heat of sifted ashes, leaving the glass open during the first twenty four hours of digestion, to evaporate the superfluous humidity. Then shut it and digest until it is become a beautiful deep red mass. This will be soon accomplished, in a heat of from 120 to 130 degrees. The trial is, that it must melt without fuming.[12]

---

[12] Here no putrefaction is mentioned by the author, perhaps there is none. -hwn

MULTIPLICATION IN QUALITY, VIRTUE AND POWER

The multiplication of the Red is performed exactly in the same manner as that of the White Tincture formerly taught.

You must dissolve the above red mass, which is the Red tincture in an infant state, capable of transmuting ten parts only of mercury into ☉, in your rectified lac Virginis, by a gentle digestion.

When perfectly dissolved distill the mercurial spirit from the Tincture until it remains an oil. This being put into a digesting globe, placed in warm ashes must be dried up again, until it becomes again a red mass.

Repeat this solution and coagulation, until it will not dry up any more, but remains a fixed ruby-red oil, which shines in the dark.

This is our Elixir Rubeum tertiae ordinis, which is capable of vitrifying a great quantity, at least one hundred parts of refined gold in the crucible, which vitrified gold can convert a greater quantity, at least one thousand parts of mercury into a red tinging cinnabar or precipitate, which, finally can transmute at least an hundred parts of mercury into fine gold.

The red Tincture is capable of being still further multiplied.

Before it has vitrified gold it is the Lapis Sophorum medicinalis universalis, the Urim and Thumim, which gives light in the dark and tinges alcohol of wine into a Ruby-red essence, wherewith you can tinge a generous, old White Austrian wine into the medicine, capable of healing and overcoming all diseases, and able to preserve life beyond the general term.

The dose of this tinged wine must be small, a few drops only, and that not too often.

TO PREPARE THE SOLAR FERMENT

Take of pure gold of 24 carats, refined with the greatest care by a faithful refiner, two ounces. Get this beat into thin leaves at a Gold beater's, one whom you can trust and who will not change your gold. You ought to get enough beat to yield you two ounces of leaves.

Dissolve the Gold leaf, one leaf after another, gradually, in your Lac Virginis, mixed with

$\overline{aa}$ good Aqua Fortis in which Aqua fortis you have previously dissolved one fourth part of its own weight of sublimed Sal ammoniac to make it become Aqua regia.

Let your double solvent, consisting of the aqua regia just mentioned and your Lac Virginis, of each an equal weight, weigh twice as much as your Gold does, that is have four ounces of solvent.

Dissolve the Gold leaves gradually, without heat and you will obtain a beautiful, transparent fiery red liquid. This is the humid calcination.

Shut the digesting glass, ♄ place it in a blood-warm water bath, to digest for eight days.

Then distill the solvent from it very carefully until there remains behind an oily liquid gold.

DIGESTION

Put the Solar oil just obtained into a digesting globe glass and set it in a water bath of a blood heat for one hundred and fifty days (five months) and the Gold will die and rot, as the Silver did before.

After Blackness is over you will obtain, in about six months time the White Mercurial sulphur of Gold, which will settle all round the globe like small pearls or the eyes of Fish.

RUBIFICATION OF THE WHITE SULPHUR OF GOLD

When you have the sign just mentioned, your White Sulphur of Gold settled round the Globe like small pearls, take your glass gently out of the water bath and place it in Ashes previously warmed over a lamp to nearly the same degree of heat as the water bath was. Then increase your heat gradually to 110, 120 and 130 degrees and the White Sulphur will change into a yellow and finally into a beautiful deep red color.

The change from the White to the Red will be accomplished in five or six weeks, ☌ you will then have in your possession the Red Spiritualized Gold or Solar Ferment, extremely fusible.

SOLUTION OF THE SOLAR FERMENT
AND
REDUCTION OF THE SAME INTO
OLEUM ☉ IS

Dissolve your Red Solar Ferment in Genuine highly rectified alcohol of wine and you will have a transparent Ruby colored solution, which no art can reduce per se into ☉ again.

This Ruby Tincture is Aurum potable per se, but not Lapis Philosophorum medicinalis, yet it is a glorious restorative and curative Medicine.
Distill the Spirit of Wine in Balneo gently from the solution, per alernbicum, until there remains behind, in your glass body, a deep Ruby red Oil of Gold, that is a solar oily looking liquid, which is the Spiritual Solar Ferment, for the composition of the Red Elixir primae, secundae et tertiae ordinis.

SOLI DEO GLORIA!
FINIS

NOTES ON THE FOREGOING PROCESS

Sigismond Bacstrom, M. D.

The Introduction I consider as containing great and valuable truths in natural science. If the Modern chemists would deign to learn and understand them, and would keep sight of them in their labors they might accomplish what they now hold to be impossible, the transmutation of one metal into another.

The Philosophical Canons (in M.S.) agree in the doctrine laid down in the Introduction to this work but they as well as Irenaeus (i. e. Doctor Winthorp) in the practical part, reject every Sophic Mercury in forma aguae, olei or butyri, admitting no other except a running mercury, or ☿ vivus, only because they succeeded therewith and not with the others, and were not Philosophers sufficient to examine the Central Elementary Powers of Nature; or because they never took the trouble to be beyond their own successful labors.

There have, however, been more Philosophers that possessed the Stone, who worked with Mercurial waters or mercurial oils, than of those who worked with Sophic running metallic mercuries.

125

This seeming disagreement therefor, while the real

agent (⚷, △, life) is centrally the same,
differing only in outward appearance, ought not to
trouble the mind, much less to perplex the Studies
of a determined and indefatigable Enquirer. Remember
what Stahl says (page 321) and which Becher had said
before him.

Basil Valentine worked long-labors via humida, and
succeeded first of all with Cold retrograded into

♅ of ☉s, and that into ⚷, ☿, ♅ ⊖.
Afterwards he worked on Hungarian vitriol, separated
a mercurial water or Spirit, a red sulphurous oil,

♅ a fixed salt, and succeeded. Afterwards he
succeeded in another way; he retrograded Iron and
Copper into a Vitriol, separated the Principles,
Sulphur Mercury, Salt -- or soul, agent, fire:

mercurial spirit, patient, ▽, and the fixt ⊖ or

▽ -- the foundation of the building, the magnet and
principle of fixation. By letting those three pass

through sufferings ♅ death, Nature regenerated
them. Having succeeded in these labors he recommends
the last mentioned way as the best. He knew nothing
of a Mercury Sophic in forma metallica currente.

If you examine this subject with care you will
easily discover the central harmony and truth of
seemingly contradictory principles. A volatile
spirit of vitriol is a mineral volatile mercury; the
succeeding ponderous oil, when concentrated into a

deep red oil is a tinging metallic ♀, or fire, or
anima, not yet maturated or fixed, and the fixed
salt is the basis of the whole -- the principle of
rest and fixation.

Paracelsus had a work upon vitriol, another upon
mercury, another with mercury and Antimony and knew
nothing of a running sophic mercury. Whenever he
passed a Druggists shop in Vienna, where it is usual
for those in that line to place a large piece of
Vitriol, Antimony or Alum, in the open shop window,
he used to always take off his hat and make a bow as
he passed the piece of Vitriol; thereby declaring
the preference he gave it. The people who were
passing thought him mad.

RECAPITULATION

The following is a brief recapitulation of this Authors process, in which, for the sake of perspicuity, a somewhat different arrangement is followed and some of his terms, where he uses a great number for the same product, are disregarded.

PREPARATION OF THE SUBLIMATE

A. Purify the Mercury by rubbing it with salt and vinegar, washing and straining it through leather. Unite the mercury with corrosive sublimate of mercury **āā**, by rubbing them together till they form a grey mass; to which add an equal weight of a mixture of nitre and roman vitriol **āā**, and rub all together till well incorporated like paste. Sublime this paste in a sand-heat, and afterwards sublime it twice more per se.--Mark this with the letter A.

B. Dissolve 1 part of A, reduced to a powder, in twice its weight of good Aqua fortis. Put the Sublimate in by little and little and keep the vessel warm in a balneum. After it has stood ten days in the balneum draw off the aqua fortis alembicum, till the residuum be like hogs lard. Then cover the bottom of the cucurbit with a lute, apply an alembic, and, in a sand heat, when the humidity is all gone, sublime the mercury into the alembic; then let all cool and in clean vessels repeat the sublimation twice more. On this put the letter B.

PREPARATION OF THE GLUTEN AQULAE

C. Put half a pound of the sublimate B in such a glass as has been described; put the belly of the bottle in sand placed sideways and sublime the contents to the upper side. Next day turn the side that now contains the sublimate down into the sand, light the fire and sublime again. Repeat this every day till, after 16 to 20 days, it refuses to ascend any more. You have then the Gluten Aguilae. Mark this with C.

PREPARATION OF THE LAC VIRGINIS

Take an other half pound of the Subl. B, and in a glass vessel slightly stopped with paper expose it to a gentle heat in a balneum. In a few days it will dissolve per se. -- Or you may let it run per deliquium on an inclined plate of glass placed over a bottle with a funnel, in a cellar: in which case the liquid must be dephlegmated. This is the Lac Virginis.

D. This must be digested in a heat of 90 degrees for ten days and then distilled in balneum.--You have now the true Lac Virginis, which the Author afterwards calls Rectified Lac Virginis. Mark this with the letter D.

Query. Will the mercurial part go over?

PREPARATION OF THE WHITE SULPHUR OF NATURE

Take 1 part of the Gluten Aquilae (C) powder it and pout on it, gradually, in a digesting glass, 2 parts by weight of the Lac Virginis (D) and lute the stopper. This is uniting the male and female--the fixed and volatile--the body and spirit--the salt and mercury, (the Sulphur is in the mercury). This is Lamspring's Dragon.

E. Digest this in a heat that you can bear with your hand for 5 months. At the end of 40 days blackness

will appear, then the transient colors, lastly Whiteness. Increase the heat 20 or 25 degrees and the matter will ascend looking like Fishes eyes, after which it will settle again and look Silver white (if it should not, the effect will be the same). This is Sulphur Naturae Album. Call this E.

PREPARATION OF THE OIL OF THE WHITE SULPHUR OF NATURE

F. Weigh the White Sulphur, put it in a glass body and pour on it twice its weight of genuine alcohol of Wine. Shut the glass with a blind alembic and digest in a blood heat for 8 days. Then apply a proper alembic and distill off the alcohol till the residuum looks like a white oil. This is the Oil of the White Sulphur naturae ex mercurio. Mark this with the letter F.

This is afterwards to be united with the Oleum Sulphuris Lunae, which is the Lunar or White ferment.

PREPARATION OF THE WHITE OR LUNAR FERMENT

Take 4, 5 or 6 ounces of pure silver in leaves, or convert your silver into a calx or Luna cornea, which edulcorate with warm water and then dry it. Put the silver to twice its weight of the rectified lac virginis (D), in a digesting glass, adding the silver gradually till the whole is dissolved.

If you be short of lac Virginis you may take nitric acid $\overline{aa}$ mixed with it. Let the liquid dissolve what it will cold, and then, in a blood-warm heat let it dissolve till saturated.

Let the solution stand in the balneum nine or ten days till the solution becomes of a green color. Then pour it, still a little warm, into a low glass body, leaving the sediment behind if there is any. Apply an alembic and distill in a balneum till the greater part of the liquid is come over and the silver remains not as a calx but like a white oily liquid.

G. Pour this into a digesting glass with a long neck, and add to it good alcohol, a few drops only at a time on account of the heat occasioned by the reaction of the two, until the alcohol stands 4 fingers breadth above the oily liquid. Shut the glass and digest in a blood heat for ten days, till a perfect union is effected. Then pour it into a glass body, apply an alembic and draw off the alcohol till the Silver remains again like an oil. In a digesting globe, one third filled, digest the above, in a blood heat, suffering the superfluous humidity to evaporate for 24 hours, after which put in the stopper and lute it. Let it digest for five months and the silver will putrefy. When the blackness is past increase the heat 20 or 30 degrees and the Peacock's tail will shew itself. After 6 weeks more the white sulphur will appear. Then raise the heat a little and the sulphur will ascend, and settle round the sides of the glass, white and shining like Fishes eyes. This is the Lunar or White Ferment. Call this G.

CONVERSION OF THE SAME INTO A WHITE LUNAR OIL

Take the White Ferment G out of the digesting globe; put it in a clean digesting glass, pour alcohol over it, two or three fingers breadth, digest
H. In a blood heat for two or three days and then, per alembicum draw off the alcohol till the Lunar ferment remains behind in the form of an oil. This is the Oleum Sulphuris Lunae. Mark it with H.

FERMENTATION OF THE WHITE MERCURIAL OIL WITH THE LUNAR OIL, FOR THE WHITE STONE

I. Take 1 part of the Oleum Sulphuris Lunae H, and add to it 3 parts of the Oleum Sulphuris Naturae Albi F. After 24 hours digestion to evaporate superfluous humidity close the glass and lute the stopper. Digest in a heat of 90 to 100 degrees. Various transient colors will pass; the matter will dry up, and become white and glittering like silver, in a few weeks. This is the Lapis Albus, but only able to go one part on ten of the baser metals. Mark this with the letter I.

MULTIPLICATION OF THE WHITE STONE IN POWER

K. Dissolve the White stone I, in the rectified lac virginis D. Distill off the lac virginis till there remains a white fixed oil. Coagulate and try this up in a heat of 100 to 120 degrees. Repeat solution[13] and siccation three or four times, until it will no more dry up but remains a fixed, incombustible oil. This is the Great White Elixir. Mark it with the letter K.

---

[13] Query. Does the Author mean that the same lac virginis which was distilled from the white stone, or fresh lac virginis is to be used for these solutions? —hwn

MULTIPLICATION OF THE GREAT WHITE ELIXIR IN QUANTITY

Take 1 part of the Great White Elixir K and project it on 100 parts of pure silver in fusion; keep them in a strong heat for at least an hour, till, on taking out a little on the end of an iron rod, you find it has become a brittle.

L. white, vitrious mass. This is the Corporeal, fermented, white metallic Elixir. Put the letter L on the bottle in which you keep it. Envelope 1 part of L in wax and put it to 100 parts of purified mercury in a crucible. Give a good heat for an hour, in a wind furnace, and the mercury M. will be converted into a fixed, fusible, white, lunar tinging precipitate. Mark this with the letter M.

PROJECTION FOR SILVER

Take 1 ounce of the tinging precipitate M, envelope
it in wax and put it to 100 ounces of common
mercury, lead, or tin; give a good heat for half an
hour and the result will be pure Silver.

If it is still brittle add a little more mercury,
lead, or tin till the whole mass becomes ductile
fine silver.

MANIPULATIONS FOR THE RED ELIXIR

The foregoing is the Process of the work for the transmutation of the baser metals into Silver. To obtain a Red Solar Tinging precipitate the Author describes other manipulations as necessary.

PREPARATION OF THE RED SULPHUR NATURAE

N. Take one of the digesting globes, containing the White sulphur naturae ex mercurii E, and, without suffering it to cool, place it in a Lamp furnace, warmed to the same degree the globe had in the bath. Continue the heat at 120 to 130 degrees till all the White Sulphur acquires a very bright cinnabar color, which it will in about a month.

This is Sulphur rubrum naturae indeterminatum. Call it N.

PREPARATION OF THE OIL OF RED SULPHUR OF NATURE

The Red sulphur N, must be brought into the form of a Red Oil (oleum rubrum sulphuris naturae) exactly in the same manner as the White sulphur naturae was brought into a White oil (See F).

0. Call this Red Oil 0; it tinges wine into a glorious medicine for the human body.
You must now proceed to—

THE PREPARATION OF THE RED OR SOLAR FERMENT

Dissolve two ounces of Gold, leaf by leaf, in a
mixture of 2 ʒ of Lac Virginis (D) with 2 ʒ of
aqua regia (made by dissolving 1 part of sal
ammoniac in 4 parts of good Aqua fortis). Dissolve
without heat. You will obtain a red solution. Shut
the glass and digest in a blood heat for 3 days--
then distill the Aqua regis ⚭ Lac virginis from it
till there remains an oily liquid gold.

Digest this in a globe glass, in a blood heat, ⚭
the gold will putrefy and show blackness in about
five months; after which, in about 6 months time,
P. the White Mercurial sulphur of gold will show
itself all round the globe like pearls.

Increase the heat to 110, 120, 130 degrees, and the
white sulphur will change to yellow and then to a
deep red color in five or six weeks. This is the
Solar ferment. It must be reduced to an oil (oleum
sulphuris solaris) by solution in alcohol and after
abstraction per alembicum till there remains a deep
ruby red oil of gold--Call it P. This is the Solar
ferment which is to be used for the—

FERMENTATION OF THE RED SULPHUR OF NATURE EX ☿II, OR THE COMPOSITION OF THE RED ELIXIR

Q. Take Oil of Red Sulphur of Nature (O) 3 parts and of the Solar Oil (P) 1 part. Manage exactly as you did the white fermentation (See I), only that now the heat must be from 120 to 130 degrees. It is not long in turning into a mass of a deep red color. When sufficiently digested it must be able to melt without fuming.

This is the Red Stone or Tincture, in an infant state, 1 part of which can only transmute 10 parts. Distinguish this by the letter Q.

EXALTATION OF THE RED ELIXIR IN POWER

This is effected in the same way as the Lapis Albus was multiplied in Power (See K). Dissolve the Red stone Q in the rectified lac virginis D, employing a gentle digestion, ⚗ then distill the lac from the Red tincture until an oil remains.

Digest this, in a globe glass, in warm ashes or sand till it becomes a red mass. Repeat the solution and coagulation till it refuses at last to dry up any more, but remains a red oil which shines in the dark. This is the Elixir Rubrum tertiae ordinis.

MULTIPLICATION OF THE RED ELIXIR IN QUANTITY

This is done in the same way as the White Stone was multiplied, with this difference only, that whereas the latter was put to fine silver, this must be put to fine gold--1 part to 100. The whole will become, in the crucible a red brittle mass, able to transmute mercury 1000 parts into a red tinging cinnabar, which again can transmute 100 parts of mercury into Gold.

Appendix A

When mixing the Oil liquor of silver with the Spirit of Wine, especially if the calcining water was composed; aqua mercurii ⚸ aqua fortis in $\overline{aa}$, I would not put the oil to the moist fire, but the fire to the oil. In making the mixture for Ether of Vitriol I find it best to add the rectified Spirit of Wine to the oil of vitriol, extending the stronger fire in the weather which is safer than a contrary mode of proceeding. Why the Author overlooks this I cannot say, as he calls his calcining ▽ a moist △. I shall show what he means by that expression.

Every corrosive in nature must of necessity be either Acid or Alkaline. When Acid it is derived from △ and ▵̶, when alkali from ▽ and ▽̶ by means of fire. Every acid in Nature is △ extended or dilated or corporified in ▵̶ and ▽. There is more △ extended in a strong concentrated than in a weaker acid. If you could deprive the smoking acid spirit of nitre of all its acid humidity, it would absolutely manifest itself in a sudden flash of △, in the act of returning to its primitive state of universality, which as fire is invisible, but as light visible.

As one element cannot act without another it is therefore impossible that the elements can ever be perfectly separated by the art of man.

△ communicates with ▽ through the medium of ◭ and thereby impregnates it, △ and ▽ being two extremes; and when △ by the means of ◭ extends itself in ▽ the universal acid is generated, every acid proceeding from △. Flame itself, maintained by the medium of ◭, is a highly concentrated acid, more active ☞ powerful than the moist corrosive fires in nitric, sulfuric and marine acids &c., but incondensable. The principle, however, in all is the same; centrally, the last mentioned acids are all, more or less, extended, determinated moist △s by means of ◭ and ▽.

The time will come when our modern Philosophers will simplify their principles as well as their minds and be obliged to return to these truths.

The Alkaline principle is likewise a corrosive but it is exactly the contrary of the acid as to its corporification. In alkali the △ is corporified and extended as well as in acid, not however by means of △ and ▽ but by means of humidity and ▽̄. For this reason every alkali (N.B. fixt) is humidity concentrated into ▽, wherein △ is corporified.

The two universal principles, as the two first manifestations of the universal agent stand thus:

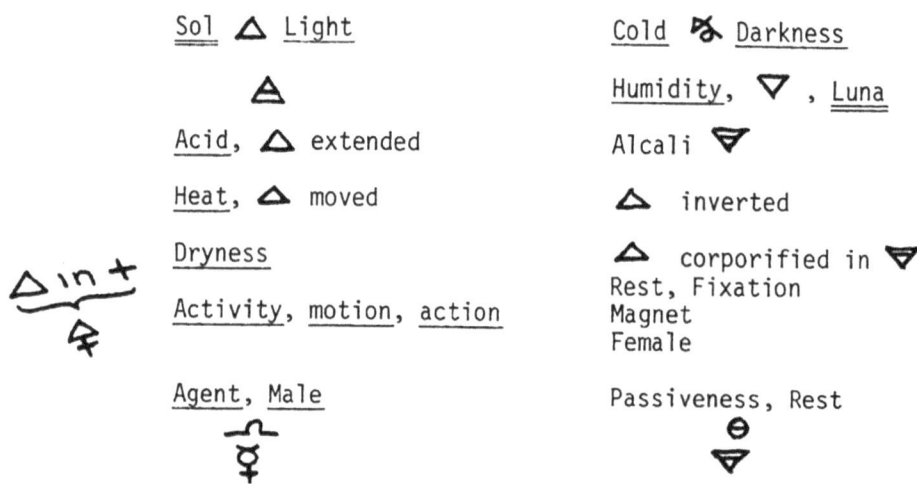

Sol △ Light	Cold ⚸ Darkness
△̄	Humidity, ▽ , Luna
Acid, △ extended	Alcali ▽̄
Heat, △̲ moved	△̲ inverted
Dryness	△̲ corporified in ▽̄
Activity, motion, action	Rest, Fixation
	Magnet
	Female
Agent, Male	Passiveness, Rest

GLOSSARY OF LATIN TERMS

Aqua fortis.	Strong water.
Aqua Mercurialis.	Water of Mercury.
Aurum potabile.	Drinkable gold.
Aqua regia.	Royal water.
B.M.	Balneum Mariae. Mary's bath (i.e. water bath).
Balneum.	Bath.
Caput Cervi.	Crow's head.
Conjunctio Spiritus cum Corpore.	The marriage of the spirit with the body.
Duplex.	Two-fold; double.
Duplicatus.	Doubled.
Elixir Rubrum Naturae indeterminatum.	The indeterminated red elixir of nature.
Elixeria tertia ordinis.	(The) Elixirs of the third order.
Ex Mercurio.	From Mercury.
Exuboratus.	Proliferated.
Fermentum Lunae.	Lunar ferment.
Finis.	The end.
Forma.	The appearance.
Fusibilis.	Fusible.
Gluten Aquilae.	Gluten of the eagle.
Ignis Gehennae.	(The) Fire of Gehenna (Hell)
In balneo	In the bath.
In forma Aquaê, olei or	In the form of water, oil

butyri.	or butter.
En forma metallica currente.	In the form of a running metal.
In forma olei.	In the form of an oil.
Indugator.	One who brings forth water; one who directs the flow of water.
Ist in mercurio quidquid quaerurit Sapientes.	Whatever the Knowing Ones seek is in (the) mercury.
Lac Virginis.	Virgin's milk.
Lapis Albus.	White Stone.
Lapis Sophorum medicinalis universalis.	The Wise Ones' stone of universal medicine.
Luna cornea.	The horned moon.
Mercurius Sophorum animatus.	The spirited Mercury of the Wise.
Natura Media.	Natural means.
Oleum Lunae.	Oil of the moon.
Oleum Solis.	Oil of the sun.
Oleum Sulphuris Lunae.	Oil of the Lunar Sulphur.
Per alembicum.	By alembic.
Per deliquium.	By removal (of sediment); by clarification, (running Into a liquid)
Per ignem suppressionis.	By the fire of concealment.
Per minima.	At least; by the least.
Per Se.	By itself.
Primae, secundae, tertiae ordinis.	Of the first, second, third order.

Quinta essentia medicinalis.	The fifth essence of medicine.
Soli Deo Gloria!	To God alone the glory!
Sulphur naturae album.	The white sulphur of nature.
Sulphur Rubrum Naturae indeterminatum.	The indeterminated red sulphur of Nature.
Via humida.	The humid way; by the humid way.
Vide.	See.
Vivus.	Living.

A Word from the Publisher

Thank you for purchasing this small work from The R.A.M.S. Library of Alchemy. During his lifetime, Hans Nintzel was dedicated to the identification, acquisition, study, retyping and, when necessary, translation of what he considered to be the most important known works on Alchemy. Hans was assisted by his sparse network of fellow Alchemists, all members of the Restorers of Alchemical Manuscripts Society (R.A.M.S.). I was an active member of R.A.M.S.

My goal is to publish all of the works originally made available through R.A.M.S. as photocopies. To facilitate this, I have chosen to have the books professionally printed. I also have a few titles that I intend to add to the original R.A.M.S. Library, selected by strict criteria established by Hans.

If you have a work on Alchemy that you believe should be a part of the R.A.M.S. Library, please contact me through R.A.M.S. Publishing Company.

Philip N. Wheeler

www.ingramcontent.com/pod-product-compliance
Lightning Source LLC
Chambersburg PA
CBHW050715180526
45159CB00003B/1034